Ronaldo Alberto
Aireen Otanes

Fungal Plant Pathogens Attacking Soursop (Annona muricata L.) Vell"

Ronaldo Alberto
Aireen Otanes

Fungal Plant Pathogens Attacking Soursop (Annona muricata L.) Vell"

LAP LAMBERT Academic Publishing

Publisher:
LAP LAMBERT Academic Publishing
is a trademark of
International Book Market Service Ltd., member of OmniScriptum Publishing Group
17 Meldrum Street, Beau Bassin 71504, Mauritius
Printed at: see last page
ISBN: 978-613-9-98176-2

ACKNOWLEDGEMENT

This research work is never the work of anyone alone. The contributions of many different people, in their different ways, have made this work possible. The author would like to extend her appreciation especially to the following:

To her parents; Mr. Warlito Otanes and Mrs. Dominga Otanes for their prayers, unconditional love, continued support over the years and faith in me;

To her sister; for her understanding, support, both emotional and financial, and assistance in numerous ways;

To Mr. Anatalio and Justina Moises for their trust, encouragement and financial support;

To her adviser, Dr. Ronaldo T. Alberto, for the supervision. His advices, patience, unfailing guidance, encouragement and immense knowledge were key motivations throughout this study;

To his critic, Mr. Zosimo G. Battad II, for his excellent feedback and suggested improvements of this study;

To Dr. Elaida R. Fiegalan, for her warm encouragement, valuable advice and comments for the completion of this work.

To the faculty and staff of the Department of Crop Protection, Dr. Roselie L. Galang, Mrs. Maribeth L. Parugrug, Mrs. Neri Santiago and Ms. Sheryl J. Marcha for their technical assistance, unparalleled kindness they have imparted to the author.

The author owe a great debt of gratitude to Dr. Annie Melinda Paz-Alberto, Director, Institute for Climate Change and Environmental Management (ICCEM), for allowing the author to work at Biodiversity Molecular Laboratory to get technical advice and assistance and to use all the necessary equipment throughout the conduct of this study.

To all the research assistants at the Molecular Laboratory, especially to Myra, Darling, Ronel, Rommel and Arnel for providing a good and friendly atmosphere in the lab, for their supervisions and sharing their knowledge and skills throughout the conduct of this study.

To Ms. Ella Joyce S. Paragas (Department of Statistics, CAS, CLSU) for her extended help in statistical analysis.

To her co- majors and friends especially to, Mea, Cathy, Jackie, Michelle, Mildred, Yza, Isip, Urma, Erwin, Dexter, April and Babylyn, for their camaraderie, untiring support and encouraging the author to do her best.

To the scholarship foundation, Crop Protection Association of the Philippines (CPAP), for the trust and financial support given to the author.

To her Abioger's family especially to Christian, Erick, Ep-Ep, Wewet, Den-Den, Jerome, Michael, Richard, Rex, Myka and Mariz for the reminiscences, emotional and spiritual support they have imparted to the author and:

Above all, thank God for the wisdom that has been bestowed upon me during this research work, and indeed, throughout my life.

To those names who were not mentioned but their encouragement and support cannot be overlooked, a heartfelt thanks is accorded them.

AIRENE T. OTANES

LIST OF TABLES

LIST OF FIGURES

FIGUREPAGE

MORPHOLOGICAL AND MOLECULAR IDENTIFICATION AND FUNGICIDE SENSITIVITY ASSAY OF PATHOGENS ATTACKING GUYABANO (*Annona muricata* L.) VELL [1]

by

AIRENE TUMBAGA OTANES

ABSTRACT

This study was conducted to characterize the plant pathogens attacking guyabano fruits and leaves through morphological and molecular approaches and to determine the fungicides where these pathogens are sensitive. Three fungal pathogens were found to be pathogenic to guyabano fruit, these pathogens are: *Colletotrichum gloeosporioides, Colletotrichum acutatum* and *Fusarium chlamydosporum. Colletotrichum gloeosporiodes* are from leaves. Molecular identification of the *Colletotrichum* species was carried out through amplification of rDNA ITS regions by using species specific primers (*Cg*Int) for *Colletotrichum gloeosporioides* and (*Ca*Int2) for *Colletotrichum acutatum* in combination with ITS4 universal primer. Ver ITS primer was used on *Fusarium* species. All fungal pathogens were found to be highly pathogenic in guyabano fruits and leaves. Among the fungicides tested, *Colletotrichum gloeosporioides, Colletotrichum acutatum* and *Fusarium chlamydosporum* were found to be highly sensitive to Captan, Tebuconazole and Difeconazole + Propiconazole.

[1] An undergraduate thesis manuscript presented in partial fulfillment of the requirements for graduation with the degree of Bachelor of Science in Agriculture major in Crop Protection from Central Luzon State University, Science City of Muñoz, Nueva Ecija. Prepared in the Department of Crop Protection under the supervision of Dr. Ronaldo T. Alberto with Research Contribution No. CA-03-14-0004.

INTRODUCTION

Background of the Study

Soursop (*Annona muricata* L.) also known as guyabano belonging to the family Annonaceae are indigenous to the tropical North and South America. It is one of the exotic fruits prized for its very pleasant, sub-acid aromatic and juicy flesh which consist of edible pulp and indigestible blade seeds. It is a major source of income for many farmers who cultivate them for fresh produce markets. It also provides source of nutrients and play important role in the diet of many people (Okigbo and Obire, 2009). The edible portion of the fruit is 70% with food energy of approximately 63 calories and sugar content ranging from 4% to 14%. The juice obtained from the fruit is used as diuretics and as a remedy for hematuria and urethritis. When drunk on an empty stomach, this juice is believed to alleviate liver diseases and leprosy. (Nweke and Ibiam, 2012). According to Tabasum (2012) the principal interest in this plant is its strong anti-cancer effects. Although it is effective for a number of medical conditions, its anti tumor effect is of most interest.

This plant is a proven cancer remedy for cancers of all types, effectively targeting and killing malignant cells in 12 types of cancer, including colon, breast, prostate, lung and pancreatic cancer. The US National Cancer Institute performed the first scientific research in 1976. The results showed that guyabano leaves and stems were found effective in attacking and destroying malignant cells. The guyabano leaves contain anti-cancer substances called annonaceous acetogenin, which can kill cancer cells without disturbing the healthy cells in the human body. Soursop leaves are said even more abundant on anti-cancer substances because soursop leaves kills only the cells that grow abnormally and let the cells grow normally (http://infonewssoftware.blogspot.com/ 2011/02/ cancer-treatment-with-soursop-leaf-html). In a study done by the Department of Science and Technology-Industrial Technology Development Institute (DOST-ITDI), it was found out that the unripe guyabano has more flavonoids than its ripe form. Flavonoids may help in preventing cancer, allergies, infections, and viruses.

According to Jaramillio (2000), the oil extracted from the unripe fruit is mixed with olive oil and used externally against neuralgias, rheumatism and arthritic pains. As stated by Okigbo and Obire (2009) national and international demand of soursop has been growing at 3.8 % per year, and the potential for developing a profitable post-harvest industry is high. However, adverse physical and chemical changes in these fruit tissues can lower acceptability and nutritional value. Generally, loss in nutritional value of fruits is brought by about by contamination by different microorganisms either before harvest or after harvest. In this crop, Joffe and Mc Donald (2001) also reported that fungal contamination is most prevalent and that fungi are their most important pathogens.

Statement of the Problem

Diseases are often the most important constraints in the production of guyabano. Fungal pathogens are known to be responsible for the post-harvest deterioration of soursop fruits (Setiawan et al., 2001). They debilitate the plants and directly reduced the yield and quality of fruits before and after they are harvested. They ranged from aesthetic problems that devastate local or regional production (Ploetz, 2003).

Gottsberger (1988) reported that the post-harvest fungal soft rot pathogens of guyabano may include the following: *Botryodiplodia theobromae, Colletotrichium gloeosoporioides, Rhizopus stolonifer, Rhizopus nigrican, Aspergillus flavus* and *Aspergillus niger.* Anthracnose generates losses up to 90% in traditionally grown soursop crops. The pathogen attacks the leaves, branches, flowers and fruits, producing black fruit rot, especially during rainy season (Alvarez et al., 2005).

Objectives of the Study

The main purpose of this work is to: (a) isolate, test for pathogenicity and identify the pathogens causing deterioration of guyabano leaves and fruit using morphological and molecular approaches; (b) determine the fungicides where these pathogens are most sensitive; (c) determine which among the pathogens is the most destructive.

Scope and Limitation of the Study

This study is limited only to the morphological and molecular identification and fungicide sensitivity assay of fungal pathogens causing deterioration of guyabano fruit and leaves and on the determination as to which among the pathogens is the most pathogenic.

REVIEW OF LITERATURE

Guyabano (*Annona muricata* L.)

The genus *Annona*, of the family Annonaceae, includes 60 or more species mostly of tropical American origin. *Annona muricata* L. is the species out of the genus *Annona* which produces biggest flowers as well as the biggest fruits. The fruits of *Annona muricata* L. are oval or heart-shaped, sometimes irregular due to bad pollination. The size of fruit is very variable in a range from 10 to 30 cm in length and 20 cm in width with a weight of 0.5 to 1.0 kg. The skin of the fruit has many short, fleshy and pointed protuberances and is popularly regarded as "spiny" (Kunz, 2007). The tree can reach heights of 30 feet. It does not tolerate cold and is susceptible to strong winds. The tree grows quickly, often producing its first crop within 3 to 5 years from seeding. The soursop is considered to be originated in the lowlands of Central America. It was first described by Spanish historian Gonzalo Fernandez de Oviedo y Valdes in 1526. The fruit is now commercially cultivated throughout the tropical Americas and is found in small family orchards in Southeast Asia, Philippines, India, Hawaii and other pacific Islands (Love and Paull, 2011). Early years of 90s found that guyabano leaves considered as cancer treatment. Attacking cancer cells with naturally safe and effective, without nausea, weight loss, hair loss, as happened in chemo therapy (http://infonewssoftware.blogspot.com//2011/02/cancer-treatment-withsoursople af .html).

Okigbo and Obire (2009) have recommended the utilization of Guyabano fruit juice for wine production with regard to a palatable and acceptable wine obtained from fermentation of Guyabano fruit juice which has a considerable high level of alcohol. Recently, Komansilan et al., (2012) found from preliminary test results that the ethanol extract of active soursop seeds has a potential larvicidal agent with an LC50 =244.27 ppm. Based on BAS crop statistics of 2003; a total land area of 3,016 has. were planted to guyabano with the following as the five leading producing regions: WesternVisayas (705 has.); Region IV-A (643 has.); Cagayan Valley (400 has.); Central Visayas (169 has.); and Central Luzon (165 has.)

Pest and Diseases

Soursop is attacked by a variety of pests depending on the region where it is located. Mealy bugs may occur in masses on the fruit, while scale and lacewings may infest the tree. The fruit maybe attacked by fruit flies and red spider mites can be a problem in dry climates (www.echonet.org). According to Morton (1987) soursop trees in very humid areas often grow well but bear only a few fruits, usually of poor quality, which are apt to rot at the tip. Most of their flowers and young fruits fall because of anthracnose caused by *Colletotrichum gloeosporioides* Penz. The same fungus causes damping-off of seedlings and die-back of twigs and branches. In the East Indies, soursop trees are sometimes afflicted with the root-fungi, *Fomes lamaoensis* Murr. and *Diplodia* sp. and by pink disease due to *Corticum salmonicolor* Berk. According to De Q. Pinto et al. (2005) *Annona* trees are attacked by a large number of insect pests and numerous diseases, although many of them are not economically important. The most important pests are aphids, mealy- and scale- bugs and fruit flies.

There are several diseases attacking the fruits of *Annona* species during the pre-and post- harvest phases: anthracnose (*Glomerella cingulata*), black cancer (*Phomopsis annonacearum*), diplodia rot (*Botryodiplodia theobromae*), purple blotch (*Phytophthora palmivora*), and brown rot (*Rhizopus stolonifer*) (De Q. Pinto et al., 2005 and Ploetz, 2003).

As stated by Okigbo and Obire (2009) different fungal species occurred within both the rotten and apparently fresh healthy fruits of *Annona muricata* L. and that some of these fungal pathogens especially *Botryodiplodia theobromae* have the potential to induce rot on fresh fruits which might have a remarkable effect on the value of fruit especially in the food industry as well as human health. The fungal pathogens responsible for the pre-harvest deterioration of soursop might have probably resided in dead stems and then dispersed by rain splash into the growing fruits to initiate infection spore, and that it was also possible that insect vectors dispersed the pathogens (Amusa et al., 2003). Fungal pathogens are known to be responsible for the post-harvest deterioration of tropical crops and *Botryodiplodia theobromae* has been reported to be of the most important fruit rot pathogen in South-western Nigeria (Adisa and Fajola, 1982). Morton (2000) also in his investigation added *Botryodiplodia theobromae* was most prevalent of the pathogens, followed by *Fusarium* spp, while *Rhizopus stolonifer* was found associated with only 7% of the fruits used in the study. Agrios (2005) observed that symptoms of post-harvest disease of *Annona muricata* L. develop during storage, but infection of fruit decay-causing pathogens could occur prior to harvest or during the post-harvest handling process and storage. Everett and Thomas (2001) stated that infection after harvest might be due to contamination of wounds by decay-causing pathogens during the drenching and packing process, and by fruit-to-fruit spread during storage. According to Bilai (1988) once a spore lands on a wounded tissue, it germinates and starts growing on the surface, producing a thick mycelium which at the same time procures cell degrading enzymes (pectinases, amylases) that denature the tissues in advance, thus leading to infection. Lunn (2004) stated that the infection process takes place in a wide temperature range of 20 to 30 °C.

Disease Incidence

Apart from pre-harvest deterioration of the fruits of Guyabano fungi, there are also some of them which cause post-harvest diseases. According to Morton (2000) potential post-harvest losses may occur. Among the important fungal pathogens are *Botryodiplodia theobromae, Colletotrichum gloeosoporioides, Rhizopus stolonifer, Rhizopus nigrican, Aspergillus flavus and Aspergillus niger* (Gottsberger, 1988).

Rhizopus spp is known to be omnipresent in the air as a contaminant, usually saprophythic, however, the infection of fleshy fruits occurs in the field through wounds or bruises (made by harvesting implements) where the tissue is in a pre- necrotic process (Kunz, 2007).

Plant Pathogenic Fungi Infecting Guyabano Fruits

Colletotrichum gloeosporioides

Colletotrichum gloeosporioides Penz. a facultative parasite belongs to the order Melanconiales. The fungus produces hyaline, one- celled, ovoid to oblong, slightly curved or dumbbell shaped conidia, 10 to 15 μm in length and 5 to 7 μm in width. Masses of conidia appear pink or salmon colored. The waxy acervuli that are produced in infected tissue are sub epidermal, typically with setae, and simple, short, erect conidiophores (Dickman, 1993).

Colletotrichum gloeosoporioides causes the disease commonly known as anthracnose on a wide range of plant species in tropical, subtropical, and temperate regions. The disease can occur on leaves, stems and fruits of Guyabano. The fungus is a primary invader of injured or weakened tissues of citrus plants in orchards and may render infected fruits un-marketable.

Unfortunately, infected tissue remains symptomless and the disease only becomes apparent after plants become senescent or have experienced stress conditions. Disease symptoms are rarely evident on the rind of fruit during the 'pre-harvest' period but at 'post-harvest' may appear as dark brown irregular lesions and that become sunken on the rind tissues. In moist conditions these lesions can also ooze pinkish spore masses from the acervuli (Hossein, 2011). The term "anthracnose" refers to the diseases caused by fungi that produce spores (conidia) in fruiting structures called acervuli (Douglas, 2011). According to Hossein (2011), *Colletotrichum gloeosporioides* colonises dead twigs and injured plant tissues and forms an abundance of acervuli and conidia. Conidia can spread over relatively short distances by rain splash or overhead irrigation. Ascospores are airborne and important in long distance dispersal. Conidia that come in contact with leaves, twigs, and fruit germinate to produce appressoria and quiescent infections which result in tissue necrosis. This tissue is subsequently colonized, acervuli are formed, thus completing the pathogen's life cycle. Dead wood and plant debris are primary sources of inocula. Fruits with quiescent infections remain asymptomatic before harvesting. Injuries and tissues weakened by other factors cause further development of quiescent infections to form lesions at post-harvesting.

Botryodiplodia theobromae

Botryodiplodia sp. belongs to the class of *Ascomycetes* and lives in a saprophytic way. They are moulds which normally need injured tissue to parasite the plant. Characteristically for this fungus is the generation of pycnidia in which the spores of the fungus are formed. Pycnidia develop on artificial media only rarely and after a long time. But if they are generated they can be seen with the naked eye as black balls of the size of pinhead. The spores are elliptical and compared to other spores relatively big. Young, immature spores are colourless and unicellular whereas mature spores are brown coloured, distichous and thick-walled. Only the presence of the mature spores allows a proper identification of the fungus (Kunz, 2007). The fast and constantly growing mycelium of the fungus is snow-white at first, turning its color within three to four weeks black. The most important species of the genus is *Botryodiplodia theobromae*.

Botryodiplodia sp. is a fungal disease attacking the fruits of *Annona* species especially in neglected orchards. Diseased fruits show symptoms of purplish to black spots or blotches confined to the surface of the fruit and eventually covered with white mycelia and black pycnidia. Diplodia rot is distinguished by its dark internal discolouration and the extensive corky rotting produces (De Q. Pinto et al., 2005).

Control of *Colletotrichum gloeosporioides*

Management and control of anthracnose involves in thorough clean up at the end of dry season, including the pruning of infected twigs, removal of rotting fruits on the ground, and then burning of all wastes. Spraying with the fungicide benomyl 0.06 %, intercalated with copper oxychloride 0.15%, every week during the rainy season and every three weeks during the dry season, gives adequate control (De Q. Pinto, 2005). In India, Singh (1992) recommended spraying with 0.05% benomyl or 0.2% mancozeb M43 at 15 to 20 day intervals.

Cultural practices to reduce disease prevalence include pruning dead wood and removal of infected plant debris to reduce dispersal of the fungus inocula; avoiding fruit injury during transport, packaging and storage process; pre-harvest insecticide application for control of fruit damaging pests and post-harvest treatment with registered fungicides (Penz and Sacc, 2011). Flowers and fruits may be affected by the anthracnose fungus and fall. This disease may be controlled by spraying the tree with fungicide such as Maneb, Captan or Vitigan blue (bpi.da.gov.ph)

MATERIALS AND METHODS

Experimental Procedure

Collection, Isolation and Maintenance of Isolates

Infected guyabano fruit and leaf samples were collected from guyabano trees in home gardens in Muñoz, Nueva Ecija. Fungal pathogens of guyabano fruits and leaves were isolated from heavily sporulating lesions following the tissue planting technique. Using a sharp scalpel, 2 to 3 mm^2 sections were cut from the advancing margin of the infected portion of the fruit and leaves. Cut sections were surface disinfected with 5.25% NaClO solution (household bleach) for 30 seconds and rinsed in 3 changes of sterile distilled water. Disinfected sections were blot dried in sterile tissue paper. Finally, four sections were planted equidistantly on a plated potato dextrose agar and incubated in darkness at 27 °C to 28 °C.

Mycelial growth arising from the planted tissues were isolated and grown into pure culture on PDA slants. The isolates were maintained through mineral oil overlay prior to refrigeration.

Table 1. Collected isolates used in the study

HOST	ISOLATE	ORIGIN
A. Guyabano fruit	*Colletotrichum* sp. 1	Magtanggol
	Colletotrichum sp. 2	Villa cuison
	Fusarium sp.	Matingkis
B. Guyabano leaves	*Colletotrichum* sp. 3	Magtanggol
	Colletotrichum sp. 4	Catalanacan

Colony and Spore Morphology

All the isolates were grown on PDA. The appearance of the colonies, the occurrence of sectors, and the vegetative and reproductive structures were described after 7 days of incubation. The colony characteristics were also described from cultures grown on potato dextrose agar plates. (Figure 1).

The cultures were incubated in darkness at 27°C to 28°C, and the diameter of the colony was recorded daily (three replicates, two measurements per replicate) for seven days.

The colony characteristics that were recorded on the seventh day were texture, color, zonation, transparency, nature of the growing margin and presence of conidial masses. For each isolate, spores were suspended in lactophenol using a sterile needle. For a more detailed observation, the slide plate culture technique has also been done. Length and width of 25 spores were measured, and conidial shape was recorded at 400× magnification using bright field microscopy. Photomicrographs of the isolates were taken.

The isolated pathogens were identified based on their colony and spore morphology and microscopic characters. These were referred to published taxonomic keys and descriptions by Deacon (2006), Pitt and Hocking (2009), Ou (1985) and Webster and Weber (2007).

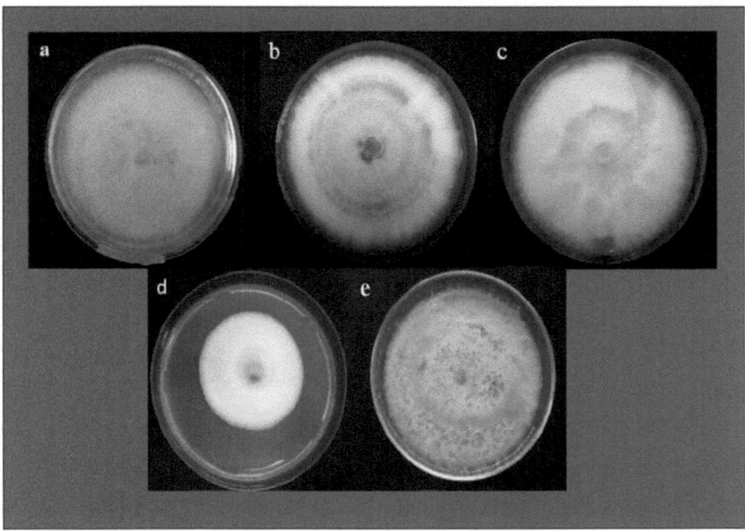

Figure 1. Fungal isolates from guyabano fruits and leaves grown on PDA for colony and spore morphology: (a) *Colletotrichum gloeosporioides* 1, (b) *Colletotrichum acutatum,* (c) *Fusarium chlamydosporum.,* (d) *Colletotrichum gloeosporioides* 2, (e) *Colletotrichum gloeosporioides* 3

PCR-Based Identification and Characterization

DNA Extraction

Isolates were grown in 100 ml potato dextrose broth for 5 days at 27°C in mechanical shaker (120 rpm). The mycelia were filtered and oven dried for 24 h at 60°C. Dried mycelia were grounded into a fine powder using liquid nitrogen and the total genomic DNA were extracted using DNeasy Plant Mini Kit (Qiagen, GmbH, Germany) following the manufacturer's protocol.

Diagnostic PCR with Species-Specific Primers

CaInt2 (5'-GGGGAAGCCTCTCGCGG-3) specific for *Colletotrichum acutatum* and CgInt (5'-GGCCTCCCGCCTCCGGGCGG-3), specific for *Colletotrichum gloeosporioides* were used in conjunction with the conserved primer ITS4. Each PCR product (50µl) contained 5 µl of forward and reverse primers, 5 µl of sterile distilled RNAse free water, 5 µl of QIAGEN™ Coral Load dye, 10 µl (mM) diluted DNA and 25 µl of QIAGEN™ mastermix. Amplification was carried out in Escoheathcare Swift Max pro Thermocycler with the following temperature profiles : 1 cycle of 5 min at 95°C, 25 cycles of 30 s at 94°C, 30 s at 62°C, and 2 min at 72°C, ending with 1 cycle of 7 min at 72°C. The PCR products (50 µl) were verified through 1.5% agarose gels containing Invitrogen™ SYBR[R] Safe DNA gel stain. The gel was checked by electrophoresis of 10 µl of reaction product in 1x TBE buffer and visualized in Maestrogen LB-16 LED transilluminator.

Sequencing of rDNA ITS Region

The amplicons (approximately 100 to 1000-bp) were purified by QIAquick[R] PCR purification kit according to manufacturer's protocols. The resulting DNA has been eluted in 60 µl of elution buffer. For each isolate, primers CaInt2 and CgInt were used for sequencing.

Purified DNAs has been sent for DNA sequencing at 1st Base Laboratories, lot 7-1 Jln SP 2/7 Tmn Serdang Perdana, Seksyen 2, 43300 Seri Kembangan, Selangor Malaysia.

Sequence Data Management and Analysis

The sequence files were assembled and edited to resolve ambiguities using sequence scanner 1.0. Databases were searched for sequence similarities using NBCI BLAST program (http://www.blast.ncbi.nlm.gov). The consensus sequence for all isolates were compiled into a single file (FASTA format) and aligned using CLUSTALW2 (http://genome.nci.nih.gov/tools/reformat.html). Phylogenetic distances were calculated by the Neighbor Joining Method (Saitou and Nei, 1987) using the Nucleotide Maximum Composite Likelihood model with the use of Molecular Evolutionary Genetics Analysis (MEGA v.4.1) software (Tamura, et al., 2007).

Pathogenicity Test

The isolated pathogens were inoculated to healthy matured guyabano fruits and leaves to confirm their pathogenicity. Fungal isolates in petri plates were grown in darkness for 14 days to induce sporulation (Figure 2). Cultures were flooded with 50 ml sterile distilled water. The surface was scraped with a sterile wire loop to dislodge the spores and was filtered through two layers of cheesecloth to remove mycelia. The spore density was adjusted to 5-6 × 10^5 spores/ml using haemocytometer. To use the haemocytometer: A cover glass was placed onto the counting chamber. A drop of inoculum was placed at the edge of the cover glass. The drop was drawn rapidly into the space between the cover glass and grid, overflow was not allowed in the moat. Cells were counted in the four corners groups and in the center group. These areas were labelled a,b,c,d and e. The number of cells were counted in all five groups and substituted in the formula:

Number of cells counted × 50 =number/mm^3×1000= number of cells/ml.

Counts were made on both counting areas twice. The dilution of a spore suspension was computed using this formula:

$$\text{Dilution factor (DF)} = \frac{\text{Current concentration}}{\text{Desired concentration}}$$

Figure 2. Fourteen (14) day- old fungal isolates grown on plated potato dextrose agar (PDA)

Fifty (50) ml spore suspension was poured on three previously pin-pricked inoculation sites marked with circles on the surface of the matured and healthy guyabano fruits and leaves. Three guyabano fruits and leaves were placed inside each of an improvised incubation chamber made up of a basin enclosed in a plastic bag. Three sterilized petri plates (bottom) were put inside the basin to serve as a base for the guyabano and to avoid water from coming into contact with the guyabano fruits and leaves. Sterilized 200 ml water was poured to enhance humidity for faster sporulation (Figures 3 and 4). Control fruits and leaves were inoculated with 50 ml of sterile water and incubated the same way as those that were inoculated with the isolates.

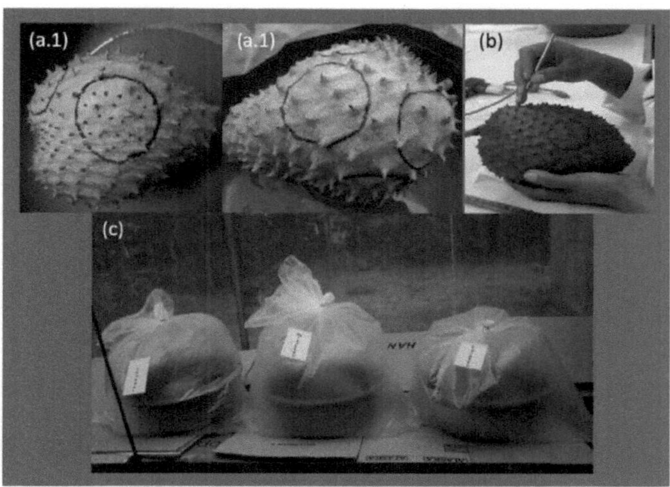

Figure 3. Pathogenecity set-up for guyabano fruits: (a.1) injured inoculated guyabano fruit on petri dish bottom as base placed with water, (a.2) uninjured inoculated guyabano fruit, (b) inoculation sites, circle marks on guyabano fruit surface being pin-pricked, (c) inoculated guyabano fruits covered with plastic bags for incubation

The set up for each pathogen was replicated three times. To satisfy the Koch's Postulates, pathogens that shows positive effects on the fruit and leaves were re-isolated and re-examined for positive identification.

Figure 4. Pathogenicity set-up for guyabano leaves: (a) inoculated guyabano leaves being sprayed, (b) inoculated guyabano leaves covered with plastic bags for incubation

Sensitivity Assay

The fungal pathogens were tested for their sensitivity to commercial fungicides based on the manufacturer's recommended rates (Table 2).

A two factor factorial experiment arranged in a Completely Randomized Design (CRD) with 12 treatments (Fungicides), using three rates, replicated three times for each isolate was used in the study. The treatments were as follows:

Table 2. Rates and treatments used in each isolates

FACTOR 1: FUNGICIDE	**FACTOR 2: RATE OF FUNGICIDES**		
	RECOMMENDED RATE	**RECOMMENDED RATE -25%**	**RECOMMENDED RATE+25%**
Tebuconazole	0.031g/50ml	0.023g/50ml	0.039g/50ml
Chlorothalonil (1)*	0.075g/50ml	0.056g/50ml	0.094g/50ml
Propineb	0.180g/50ml	0.135g/50ml	0.225g/50ml
Benomyl	0.047g/50ml	0.035g/50ml	0.059g/50ml
Cupric hydroxide	0.125g/50ml	0.094g/50ml	0.156g/50ml
Mancozeb	0.141g/50ml	0.105g/50ml	0.176g/50ml
Chlorothalonil (2)*	0.075g/50ml	0.056g/50ml	0.094g/50ml
DP	0.094g/50ml	0.070g/50ml	0.117g/50ml
Captan	0.188g/50ml	0.141g/50ml	0.234g/50ml
Azoxystrobin	0.047g/50ml	0.035g/50ml	0.059g/50ml
Carbendazim	0.047g/50ml	0.035g/50ml	0.059g/50ml
Thiophanate methyl	0.031g/50ml	0.023g/50ml	0.039g/50ml

*Chlorothalonil (1) distributed by Integrated Crop Trading Corp.
*Chlorothalonil (2) distributed by Bayer CropSci.

A 100µl of conidial suspension ($5\text{-}6\times10^5$ ml^{-1}) of each isolate was poured separately in Petri plates. After which, PDA was poured on the Petri plates, and swirled gently for an even distribution of the conidial suspension. Upon solidification, a cork borer, immersed in an Erlenmeyer flask containing ethanol was sterilized by flaming it in an alcohol lamp. After the flame has extinguished, the cork borer was allowed to cool for about 20 seconds. With a sterile cork borer held upright, the plated potato dextrose agar with conidial suspension was stabbed all the way down to the bottom of the dish, creating 5 wells (4 equidistant wells on the side and 1 on the center to serve as the control) cleanly cut from the plate. After creating 5 cups on each plate, the agar plugs inside the cork borer was removed by pushing them out with the help of a metal rod, and the agar plugs must be placed in a beaker with 10% NaClO solution to disinfect before disposal. Should the agar plug remained in the dish, use a flame sterilized teasing needle to gently scoop out the agar plug from the well. Using a micropipette, 20 µl of the fungicide treatment was dropped on each 4 wells. For the control, sterile-distilled water was dropped on the fifth well located at the center of the plate. The plates were incubated for 3 days at 28°C and the mean radius of the inhibition zone was measured across two perpendicular axes (minus the diameter of the well) for each fungicide concentration evaluated. With the ruler, the zones of inhibition were measured from the underside of the plate. Three replications of each isolate for each fungicide concentration was taken. The data was evaluated using two-factorial Completely Randomized Design (CRD) and analyzed through ANOVA using SAS v.9.1 program.

Data Gathered

Study 1 (Pathogenicity Test)

Symptom development was recorded and evaluated 72 hours after inoculation using the following parameters:

$$\textbf{(a) Percent Disease Incidence (\% DI)} = \frac{\text{Total no. of infected fruits}}{\text{Total no. of fruit samples}} \times 100$$

$$\textbf{(b) Percent Disease Severity (\% DS)} = \frac{n(0) + n(1) + n(2) + n(3) + n(4) + n(5)}{N \times 5} \times 100$$

Where:

n = no. of infected fruits classified based on rating scale

N = total number of samples

Disease severity rating was rated from each fruit according to the scale below (Ilag and Quimio, 2006):

SCALE	FRUIT RATING
0	no infected portion
1	less than 1 % of commodity surface infected
2	1-5 % of commodity surface infected
3	6-10% of commodity surface infected
4	11-20% of commodity surface infected
5	over 20% of commodity surface infected

Study 2 (Sensitivity Assay)

1. Diameter (mm) of the zone of inhibition 3 days after incubation.

RESULTS AND DISCUSSION

Morphological Characteristics of Pathogens Attacking Guyabano Fruits

Colletotrichum gloeosporioides (1)

Circular, dark sunken anthracnose lesions were observed from the guyabano fruits collected from Magtanggol. The circular spot sometimes coalesces to form irregular spots (Figure 5a).

On PDA, colonies, 66 mm diameter, cottony, dense, initially white or cream white to pale gray mycelium with a few conidial masses was observed (Figure 5b).

Microscopic examination revealed ovoid to oblong and rounded on both ends, slightly curved dumbbell shaped conidia 13.41 µm in length and 5 µm in width (Figure 5c).

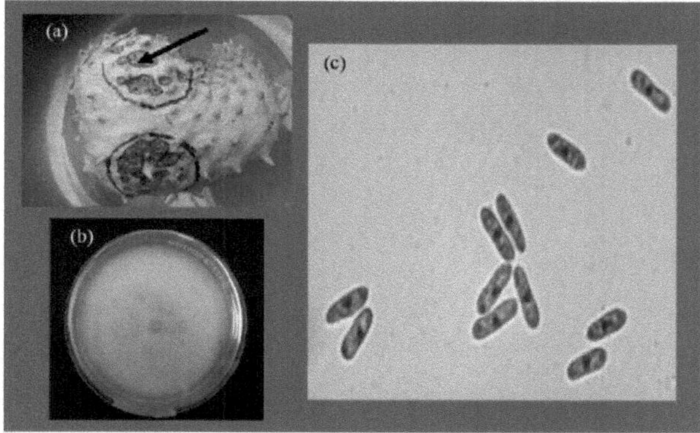

Figure 5. *Colletotrichum gloeosporioides* (1): (a) growing on the surface of a guyabano fruit, (b) 14 day old culture and (c) photomicrograph showing conidia of a 14 day old culture (400x)

Colletotrichum acutatum

Growth observed from the guyabano fruit collected from Villa Cuison begins as small, dark brown necrotic and sunken lesions and this sunken spot continues to enlarge as the fruits ripens (Figure 6a).

Colonies on PDA 67.33 mm were dense, thick, and had a fluffy textured. The cultures had cottony, white to orange mycelium with orange conidial masses produced inconcentric rings on the colony (Figure 6b).

Photomicrographs revealed 13.18 µm in length and 2.84 µm in width conidia that formed cylindrical with rounded ends but sometimes slight fusiform (Figure 6c).

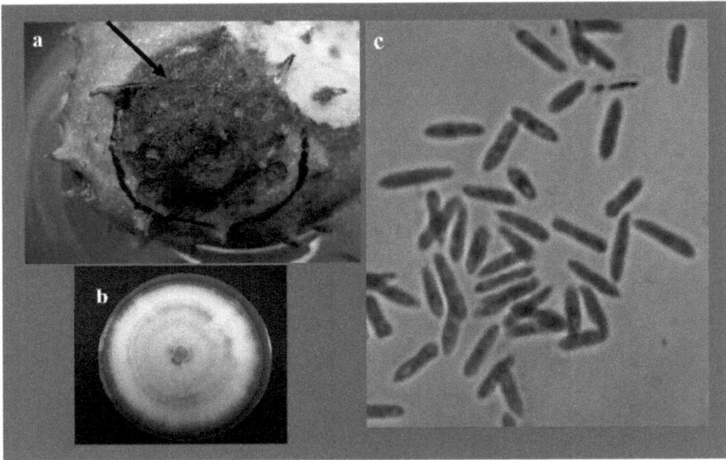

Figure 6. *Colletotrichum acutatum*: (a) growing on the surface of guyabano fruit, (b) 14 day old culture, (c) photomicrograph of conidia of a 14 day old culture (400x)

Fusarium chlamydosporum

Black necrotic slightly sunken lesion was observed on the surface of the guyabano fruit. Necrotic lesion appeared as white to greyish on the injured part of the fruit surrounded by black necrotic lesion (Figure 6a).

Colonies on PDA 57.5 mm diameter white to yellow mycelium turns fluffy, dense and cottony (Figure 6b).

Microscopic examination revealed slightly curved, fusiform, pointed at the tip mostly 3-4 septation macro conidia 20.62 µm in length and 3.88 µm in width (Figure 6c).

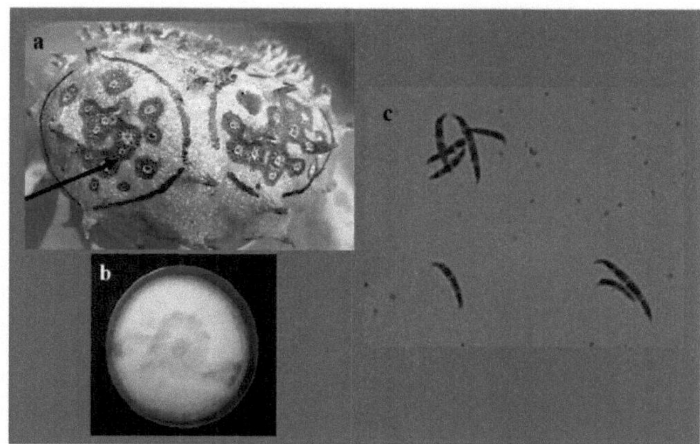

Figure 7. *Fusarium chlamydosporum*: (a) growing on the surface of guyabano fruit, (b) 14 day old culture, (c) photomicrograph of conidia of a 14 day old culture (400x)

Morphological Characteristics of Pathogens Attacking Guyabano Leaves

Colletorichum gloeosporioides (2)

Symptoms on guyabano leaves collected from Magtanggol appeared as brown angular lesion and paper-thin (Figure 7a.1).

On PDA, colonies 63 mm diameter were dense, cottony, initially white turns to gray mycelium produced orange conidial masses at the center (Figure 7b).

Conidia 14.38 µm in length and 4.32 µm in width cylindrical with rounded ends (Figure 7c).

Figure 8. *Colletotrichum gloeosporioides* (2): (a.1) growing on the surface of the leaves, (a.2) close-up of leaf angular lesion, (b) 14 day old culture, (c) photomicrograph showing conidia of a 14 day old culture (400x)

Colletotrichum gloeosporioides (3)

The leaves collected from Catalanacan showed symptoms as large, brown papery patches lesion on the surface of the guyabano leaves (Figure 8a).

Colonies on PDA 52 mm diameter appeared initially white or cream white becoming pale gray with dark or orange at the center, dense, cottony, thick and fluffy (Figure 8b).

Microscopic examination revealed cylindrical conidia with obtuse ends 14.8 µm in length and 4.1 µm in width (Figure 8d).

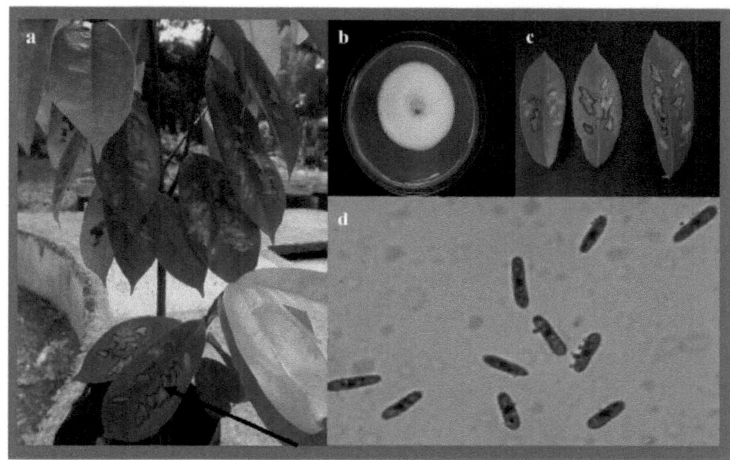

Figure 9. *Colletotrichum gloeosporioides* (3): (a) growing on the suface of the leaves (b) 14 day old culture, (c) close-up of anthracnose lesion, (d) photomicrograph showing conidia from 14 day old culture (400x)

Morphological identification of guyabano isolates was based on phenotypic traits such as colony appearance, and character of vegetative and reproductive structures. The results of cultures studies were almost fully consistent with description of Smith and Black (1990). In *Colletotrichum* isolates, the conidia of *Colletotrichum acutatum* are elliptic-fusiform in shape; whereas conidia of *Colletotrichum gloeosporioides* are cylindrical with obtuse ends (Dyko and Mordue, 1979; Baxter et.al., 1983; Smith and Black, 1990). However, a definite identification of *Colletotrichum* species based on morphology is difficult because isolates have overlapping ranges of conidial and colony characteristics, and because the variation in morphology is accepted for isolates within a species (Sutton, 1992). *Fusarium* species emphasizes the need for accurate identification on the species level. According to Leslie and Summerell (2006), *Fusarium* species determined can be identified based on morphology alone. However, identification based on molecular data is considered more reliable and accurate than morphological identification and has become much more important in diagnostics of the fungi from the genus *Fusarium*.

Pathogenicity Test

Disease Incidence (%) in Guyabano Fruits and Leaves

Symptoms developed on the guyabano fruit 72 hours after inoculation. Symptoms of rotting was very evident in all inoculated areas. All the inoculated isolates were found to be highly pathogenic to the guyabano fruits (100% disease incidence) as compared to guyabano fruits inoculated with sterile distilled water (Table 4). For the uninjured fruits, no changes and no mycelial growth was observed.

Disease incidence on guyabano leaves was observe as early as 24 hours after inoculation. Symptoms of brown papery lesions were very evident on inoculated leaves. Control leaves inoculated with sterile distilled water remain uninfected (Table 5).

Table 4. Disease Incidence (%) of guyabano fruits 3 days after inoculation

ISOLATE	ISOLATED FUNGAL PATHOGEN	%DI
1	*Colletotrichum gloeosporoiodes* (1)	100.00[a]
2	*Colletotrichum acutatum*	100.00[a]
3	*Fusarium chlamydosporum.*	100.00[a]
Control	Sterilized distilled water	0.00[b]

Table 5. Disease Incidence (%) of guyabano leaves 3 days after inoculation

ISOLATE	ISOLATED FUNGAL PATHOGEN	%DI
1	*Colletotrichum gloeosporoiodes* (2)	100.00[a]
2	*Colletotrichum gloeosporioides* (3)	100.00[a]
Control	Sterilized distilled water	0.00[b]

Disease Severity (%) in Guyabano Fruits and Leaves

At three days after inoculation, the severity of fruit deterioration has reached as high as 88% in fruit that was inoculated with *Colletotrichum acutatum*. Fruits that were inoculated with *Colletotrichum gloeosporioides* (1) and *Fusarium chlamydosporum* have reached severity of 82% and 86% respectively. Control fruits inoculated with sterile distilled water still remain unaffected (Table 6).

Meanwhile, leaves from Magtanggol inoculated with *Colletotrichum gloeosporioides* (2) have reached a severity of 59% and leaves from Catalanacan inoculated with *Colletotrichum gloeosporioides* (3) on the other hand reached a severity as high as 77%. While for the control leaves remain unaffected (Table 7).

Table 6. Disease Severity (%) of guyabano fruits 3 days after inoculation

ISOLATE	ISOLATED FUNGAL PATHOGEN	%DS
1	*Colletotrichum gloeosporoiodes* (1)	82.22[a]
2	*Colletotrichum acutatum*	88.88[a]
3	*Fusarium chlamydosporum*	86.66[a]
Control	Sterilized distilled water	0.00[b]

Table 7. Disease Severity (%) of guyabano leaves 3 days after inoculation

ISOLATE	ISOLATED FUNGAL PATHOGEN	%DS
1	*Colletotrichum gloeosporoiodes* (2)	59.89[b]
2	*Colletotrichum gloeosporioides* (3)	77.78[a]
Control	Sterilized distilled water	0.00[c]

Results of pathogenicity tests shows that *Colletotrichum gloeosporioides*-1 *Colletotrichum acutatum* and *Fusarium chlamydosporum* caused symptoms of deterioration in guyabano fruits. *Colletotrichum gloeosporioides*-2 and *Colletotrichum gloeosporioides*-3 were also proven to be the causal agent of anthracnose in the leaves. Pathogenicity test also shows that all isolates were highly virulent to cause disease in guyabano fruit and leaves that requires sensitivity test to fungicides.

Incidence and severity are the tools for measuring the diseases. According to the study of Onyeani et al. (2012) on fruit anthracnose in mango, they observed that although anthracnose was prevalent in all mango growing areas surveyed, the occurrence and severity was probably more influenced by environmental conditions and cultural practices rather than climatic factors in the areas. Fruit yield was lowest in areas with highest anthracnose disease occurrence and severity an indication that anthracnose disease phenomenon has negative correlation with fruit yield. This phenomenon was largely influenced by environmental conditions and cultural practices.

Based from the study of Ritchie (2004), on brown rot of stone fruits, the amount of inoculum is very important in determining the severity of brown rot. Warm, wet or humid weather during the 2 to 3 week period prior to harvest increases disease severity because it increases both the level of inoculum and the amount of infection. According to Hossain et al. 2010 on their survey on major diseases of fruits and vegetables, the highest disease incidence was recorded from soft rot of potato. The lowest fruit disease incidence was recorded from anthracnose of banana and papaya. They concluded that the incidence and severity of vegetables and fruits can differ in the different locations because of the existence of strains of the pathogen.

Sensitivity Assay

Colletotrichum gloeosporioides (1)

Difeconazole + Propiconazole and Captan were found to be highly effective against *Colletotrichum gloeosporioides* (1) as shown by big zone of inhibitions ranging from 19.94 mm to 21.00 mm and 12.73 mm to 14.58 mm respectively 3 days after incubation in all concentrations. This was followed by Tebuconazole that showed sensitivity in the two concentrations but to a lesser degree in recommended rate less 25%. It also showed sensitivity to Propineb at recommended rate +25% but with minimal effects on the other two concentrations (Figure 9).

Cupric Hydroxide and Thiophanate Methyl on the other hand were less effective against *Colletotrichum gloeosporioides* all rates of the fungicides tested (Table 8).

Table 8. Zone of inhibition of the 12 fungicides tested against *Colletotrichum gloeosporioides* (1) in guyabano fruit three days after incubation

FUNGICIDE	ZONE OF INHIBITION (mm)		
	Recommended Rate	Recommended Rate – 25%	Recommended Rate + 25%
1. Tebuconazole	12.60^b	7.00^{cd}	9.83^b
2. Chlorothalonil (1)	1.13^{de}	3.90^{cdef}	4.69^c
3. Propineb	6.02^c	3.17^{def}	10.48^b
4. Benomyl	2.17^{de}	7.27^c	5.40^c
5. Cupric hydroxide	0.00^e	0.00^f	0.00^d
6. Mancozeb	3.33^d	3.25^{def}	2.73^{cd}
7. Chlorothalonil (2)	7.79^c	4.27^{cde}	5.67^c
8. DP*	20.88^a	19.94^a	21.00^a
9. Captan	14.58^b	12.86^b	12.73^b
10. Azoxystrobin	1.88^{de}	0.29^f	0.96^d
11. Carbendazim	0.27^e	1.00^{ef}	0.79^d
12. Thiophanate Methyl	0.00^e	0.00^f	0.00^d
13. Control	0.00^e	0.00^f	0.00^d

*DP= Difeconazole + Propiconazole

Means with a common letter are not significantly different

The three best fungicides against *Colletotrichum gloeosporioides* (1) were Captan, Difeconazole + Propiconazole and Tebuconazole. Based from the results, the Recommended Rate is the best to use because the pathogens shows high sensitivity to this rate and has no difference with the effects in using Recommended Rate + 25% when it comes to effectiveness of the fungicides against *Colletotrichum gloeoeporioides*.

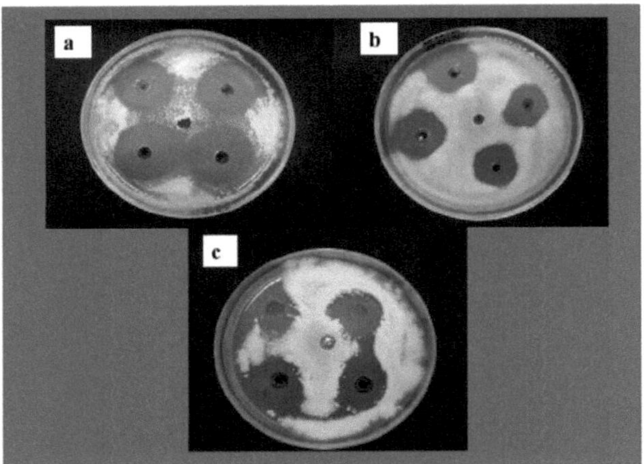

Figure 10. Zone of inhibition of the three best fungicides against *Colletotrichum gloeosporioides*: (a) Difeconazole + Propiconazole, (b) Captan, (c) Tebuconazole

Colletotrichum acutatum

Based from the results, it's in Captan that *Colletotrichum acutatum* showed the most prominent inhibition zone ranging from 14.54 mm- 18.81 mm in diameter (Table 9). *Colletotrichum acutatum* was also recorded as highly sensitive to Difeconazole + Propiconazole in all concentrations whereby the zone of inhibitions was very wide that ranges 14.58 mm to 16.08 mm in diameter (Figure 10). The rest of the fungicides tested shows minimal effects against *Colletotrichum acutatum*.

Table 9. Zone of inhibition (mm) of the 12 fungicides tested against *Colletotrichum acutatum* in guyabano fruit three days after incubation

FUNGICIDE	ZONE OF INHIBITION (mm)		
	Recommended Rate	Recommended Rate − 25%	Recommended Rate + 25%
1. Tebuconazole	0.42f	0.17c	0.71e
2. Chlorothalonil (1)	3.71d	3.46b	6.83c
3. Propineb	1.08f	2.04c	0.00e
4. Benomyl	0.00f	1.04c	0.00e
5. Cupric hydroxide	0.00f	0.00c	0.00e
6. Mancozeb	2.50ef	0.00c	3.06d
7. Chlorothalonil (2)	3.15de	1.42bc	3.04d
8. DP*	14.58b	16.08a	15.00b
9. Captan	16.67a	14.54a	18.81a
10. Azoxystrobin	7.50c	1.96bc	3.71d
11. Carbendazim	0.58f	0.29c	0.25e
12. Thiophanate methyl	0.50f	0.33c	0.67e
13. Control	0.00f	0.00c	0.00e

* DP= Difeconazole + Propiconazole

Means with a common letter are not significantly different

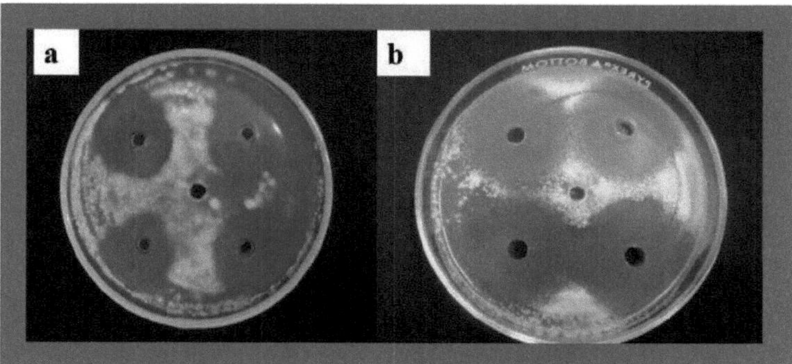

Figure 11. Zone of inhibition of the two best fungicides against *Colletotrichum acutatum* isolated from guyabano fruits: (a) Difeconazole + Propiconazole, (b) Captan

Fusarium chlamydosporum

The biggest inhibition zone recorded against *Fusarium chlamydosporum* was on Difeconazole + Propiconazole (16 mm − 18.23 mm) and sensitivity of the isolate in all concentrations was also seen. The results also show Captan as highly effective as with

Difeconazole + Propiconazole with the zone of inhibition of *Fusarium chlamydosporum* ranging from 10.69 mm up to 17.98 mm diameter. Moreover, Azoxystrobin shows its efficacy in using the recommended rate +25% with its zone of inhibition of 9.25 mm diameter. *Fusarium chlamydosporum* shows no sensitivity against Tebuconazole, Benomyl, Cupric Hydroxide and Thiophanate Methyl and less sensitive on the rest of the fungicides tested (Table 10). The two best fungicides are shown on Figure 11.

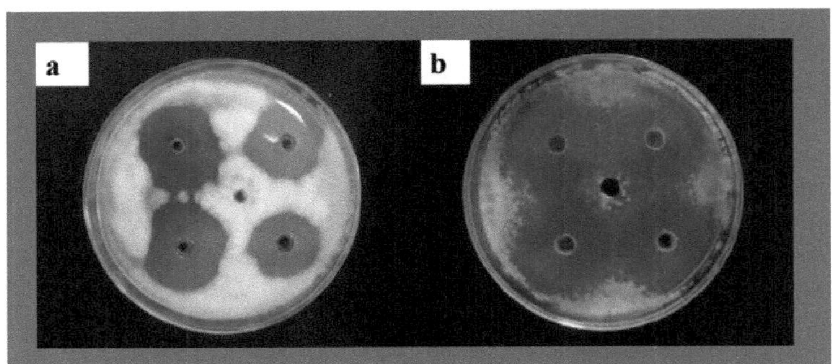

Figure 12. Zone of inhibition of the two best fungicides against *Fusarium chlamydosporum* isolated from guyabano fruits: (a) Captan, (b) Difeconazole + Propiconazole

Table 10. Zone of inhibition (mm) of the 12 fungicides tested against *Fusarium chlamydosporum* in guyabano fruits three days after incubation

FUNGICIDE	ZONE OF INHIBITION (mm)		
	Recommended Rate	Recommended Rate - 25%	Recommended Rate 25%
1. Tebuconazole	0.00[f]	0.00[e]	0.00[c]
2. Chlorothalonil 1	2.19d[e]	1.17[de]	0.00[c]
3. Propineb	7.29[b]	6.33[c]	2.75[c]
4. Benomyl	0.00[f]	0.00[e]	0.00[c]
5. Cupric Hydroxide	0.00[f]	0.00[e]	0.00[c]
6. Mancozeb	0.13[f]	0.60[de]	2.81[c]
7. Chlorothalonil 2	1.00[ef]	0.58[de]	0.35[c]
8. DP*	16.25[a]	18.23[a]	16.00[a]
9. Captan	17.98[a]	10.69[b]	16.12[a]
10. Azoxystrobin	3.21[d]	7.88[bc]	9.15[b]
11. Carbendazim	5.15[c]	4.44[cd]	7.21[b]
12. Thiophanate Methyl	0.00[f]	0.00[e]	0.00[c]
13. Control	0.00[f]	0.00[e]	0.00[c]

*DP= Difeconazole + Propiconazole

Means with a common letter are not significantly different

Colletotrichum gloeosporioides (2)

Sensitivity of *Colletotrichum gloeosporioides* was observed on Difeconazole + Propiconazole (14.92 mm – 20.02 mm) in all concentrations with the highest zone of inhibitions recorded in the recommended rate +25% (20.02 mm diameter). Captan was also found to be highly effective at recommended rate +25% with 19.63 mm diameter zone of inhibition. Thiophanate Methyl and Mancozeb were observed to be ineffective. Minimal effects were observe on the rest of the fungicides tested (Table 11.) The two best fungicides are shown on Figure12.

Table 11. Zone of inhibition (mm) of the 12 fungicides tested against *Colletotrichum gloeosporioides* (2) in guyabano leaves collected from Magtanggol three days after incubation

FUNGICIDE	ZONE OF INHIBITION (mm)		
	Recommended Rate	Recommended Rate - 25%	Recommended Rate 25%
1. Tebuconazole	1.54^{cde}	4.60^c	1.04^c
2. Chlorothalonil 1	3.65^{bc}	3.38^c	1.19^c
3. Propineb	2.08^{cde}	2.58^{cd}	4.02^b
4. Benomyl	1.21^{de}	0.63^{def}	1.21^c
5. Cupric Hydroxide	0.54^e	0.88^{def}	0.33^c
6. Mancozeb	0.00^e	3.27^c	3.52^b
7. Chlorothalonil 2	3.29^{bcd}	3.17^c	4.69^b
8. DP*	16.92^a	14.92^a	20.02^a
9. Captan	15.15^a	13.83^a	19.63^a
10. Azoxystrobin	3.27^{bcd}	2.38^{cde}	3.21^b
11. Carbendazim	0.79^e	0.27^{ef}	0.40^c
12. Thiophanate Methyl	4.75^b	8.21^b	0.00^c
13. Control	0.00^e	0.00^f	0.00^c

*DP= Difeconazole + Propiconazole

Means with a common letter are not significantly different

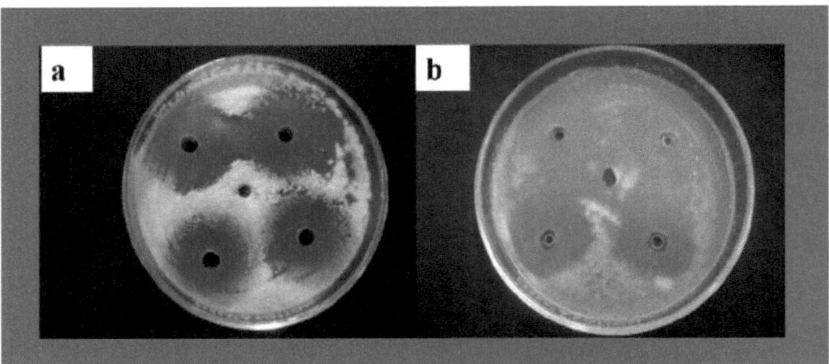

Figure 13. Zone of inhibition of the two best fungicides against *Colletotrichum gloeosporioides* isolated from guyabano leaves: (a) Difeconazole + Propiconazole, (b) Captan

Colletotrichum gloeosporioides (3)

Most prominent inhibition zone was recorded in Difeconazole + Propiconazole with zone of inhibition ranging from 19.73 mm up to 20.30 mm diameter and the widest was observe in recommended rate +25% (20.30 mm). Captan was also found to be highly effective in all concentrations (14.75 mm- 18.10) (Figure 14.). On the other hand, Thiophanate Methyl was observed to have no effects in all concentrations as well as in the Cupric Hydroxide (Table 12).

Table 12. Zone of inhibition (mm) of the 12 fungicides tested against *Colletotrichum gloeosporioides* (3) in guyabano leaves from Catalanacan three days after incubation

	ZONE OF INHIBITION (mm)		
FUNGICIDE	**Recommended Rate**	**Recommended Rate - 25%**	**Recommended Rate 25%**
1. Tebuconazole	3.17^d	1.56^{de}	5.60^b
2. Chlorothalonil 1	1.65^{de}	0.00^e	2.75^{cd}
3. Propineb	2.21^{de}	2.48^d	3.40^{bc}
4. Benomyl	6.46^c	0.17^e	2.86^{cd}
5. Cupric Hydroxide	0.00^e	0.00^e	1.25^{cde}
6. Mancozeb	0.81^{de}	0.46^e	1.40^{cde}
7. Chlorothalonil 2	3.12^d	4.52^c	2.42^{cde}
8. DP*	20.17^a	19.73^a	20.30^a
9. Captan	16.58^b	14.75^b	18.10^a
10. Azoxystrobin	0.18^e	4.33^c	3.04^{cd}
11. Carbendazim	0.54^{de}	0.42^e	0.63^{de}
12. Thiophanate Methyl	0.00^e	0.00^e	0.00^e
13. Control	0.00^e	0.00^e	0.00^e

*DP= Difeconazole + Propiconazole

Means with a common letter are not significantly different

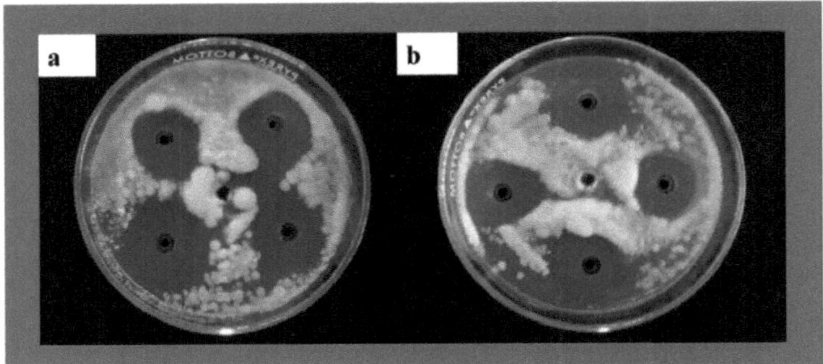

Figure 14. Zone of inhibition of the two best fungicides against *Colletotrichum gloeosporioides* isolated from guyabano leaves collected from Catalanacan: (a) Captan, (b) Difeconazole + Propiconazole

Difeconazole + Propiconazole and Captan were the most effective against *Fusarium chlamydosporum*, *Colletotrichum gloeosporiodes* and *Colletotrichum acutatum*. Different fungicides have different mode of action against different pathogens and Captan belongs to curative broad spectrum fungicides, class phenylthalamides and degrades to thiophosgene, a compound that inhibits a number of fungal enzymes. Captan is a non-specific thiol reactant with protective and curative action that works by inhibiting respiration of numerous species of fungi including: several species of *Fusarium* and *Colletotrichum* sp. and bacteria (Mueller and Bradley, 2010). Difeconazole + Propiconazole, on the other hand, has a broad, excellent activity against all pathogens classes (Dahmen and Staub, 1992). The rest of the fungicides were less effective to the tests organisms because of their limited and specific spectrum against the test pathogen

Agar-well diffusion technique has been widely used to determine the fungicides resistance in populations of fungal plant pathogens. Bioassay determine dose response relationships, and compare antifungal activity with fungicides of known mode of action (Wedge, 2005).

According to Hooper et al. (2008), the agar diffusion assay is one method for quantifying the ability of antibiotics to inhibit bacterial growth. The agar diffusion assay is an important technique for assessing microbial susceptibility to antibiotics, which has found application worldwide over the past 50 years. It involves the application of antibiotic solutions of different concentrations to cups, wells or paper discs, placed on the surface of or punched into agar plates seeded with the test bacterial strain. Antibiotic diffusion from these sources into the agar medium leads to inhibition of bacterial growth in the vicinity of the source and to the formation of clear 'zones' without bacterial lawn. The diameter of these zones increases with antibiotic concentration.

Molecular Characterization of Pathogens Attacking Guyabano

DNA extraction was done by growing the isolates in Potato Dextrose Broth for five days in orbital shaker to increase oxygenation of mycelia (Figures 15 and 16).

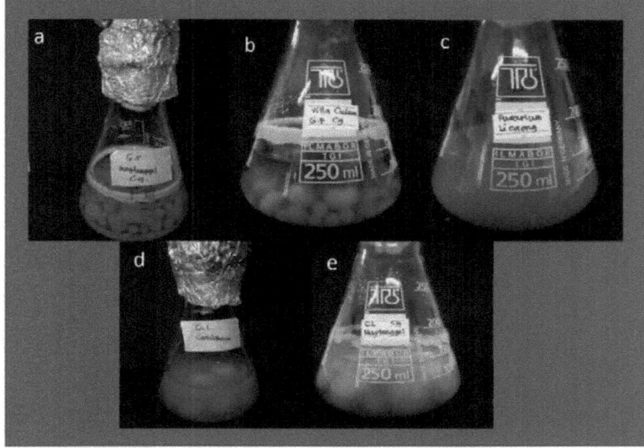

Figure 15. Isolates from guyabano fruits and leaves grown on PDB: (a) *Colletotrichum gloeosporioides* (1), (b) *Colletotrichum acutatum*, (c) *Fusarium chlamydosporum*, (d) *Colletotrichum gloeosporioides* (2), (e) *Colletotrichum gloeosporioides* (3)

Figure 16. Isolates from guyabano fruits and leaves grown on PDB in a mechanical shaker for 7 days

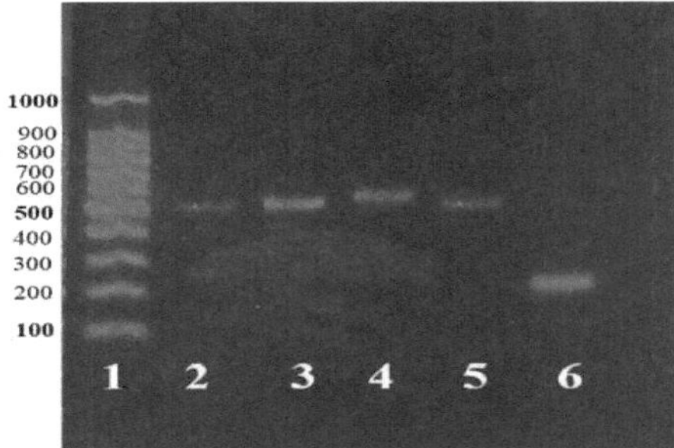

Figure 17. Amplification of DNA fragments from *Colletotrichum* and *Fusarium* isolates: Lanes 2 and 5; *C. gloeosporioides* from guyabano leaves, Lanes 3 and 4; *C. gloeosporioides* and *C. acutatum* from guyabano fruits and, Lane 6; *Fusarium chlamydosporum* isolate from guyabano fruit. 1; Molecular DNA Ladder (100-1000-bp)

The species-specific primer *Ca*Int2 for *Colletotrichum acutatum* and *Cg*Int for *Colletotrichum gloeosporioides* in conjunction with ITS4 primer amplified a 490-bp and 450-bp fragments from genomic DNA of *Colletotrichum gloeosporioides* and *Colletotrichum acutatum* of guyabano fruits and leaves respectively confirming the species identity. Universal primer Ver ITS for *Fusarium chlamydosporum* amplified a 250-bp fragment (Figure 17).

Phylogenetic Analysis

The aligned sequence of *Colletotrichum gloeosporioides* isolated from guyabano fruit when compared to different *Colletotrichum* species in GenBank were found 99% similarity to the identity of *C. gloeosporioides* in GenBank with accession number JX902437.1. Phylogenetic analysis showed that the *Colletotrichum* isolate was phylogenetically related to *C. gloeosporiodes* present in database. The level of confidence of the NJ tree was well supported with 98% bootstrap value clustered to the main clades of *C. gloeosporioides* (JX902437.1, KC209100.1, KJ152170.1, KC122768.1 and KF923853.1), *Colletotrichum fragariae* (KC209101.1) and *Glomerella cingulata* (KF923855.1) and clustered separately to the *Phytophthora capsici* as an out group from the phylogenetic tree (Figure 18). On the other hand, *Colletotrichum acutatum* in guyabano fruit isolates were matched to *Glomerella acutata* with 99% similar identity present in data base. The *C. acutatum* isolates also clustered to the clades of *G. acutata* with accession number DQ454010.1 based on NJ tree analysis supported with 99% bootstrap value (Figure 19). Moreover, another fungal pathogen, *Fusarium chlamydosporum* generated sequences with 99% similarity to *F. chlamydosporum* (KF494035.1) in the database. Figure 20 showed the clades supported with 89% bootstrap value of *F. chlamydosporum* isolates that clustered to the group of *F. chlamydosporum* (KF494035.1, KF494034.1 and KF494033.1) and *Fusarium solani* (KF918580.1). Meanwhile, fungal pathogen of guyabano leaves from Magtanggol and Catalanacan were also isolated. Analysis also shows that both isolates were identified as *C. gloeosporioides* coded as isolate 2 and 3 respectively. The *C. gloeosporioides* 2 was closely related to *C. gloeosporioides* JF487788.1 and clustered within the group of *C. gloeosporioides* with 99% bootstrap value (KJ168597.1, JX258684.1, KC010537.1, KC565728.1 and KF923871.1) and *Glomerella cingulata* DQ062671.1 which separated from *Tilletia barclayana* (AY837518.1) as out group (Figure 21). While the identified *C. gloeosporioides* 3 was closest to *C. gloeosporioides* (KC209102.1) with 99% similarity in the database. The clades of the tree supported with 100% bootstrap value includes *C. gloeosporioides* (KC209102.1, KF907247.1, HQ328013.1, KF923871.1, KJ152170.1 and JX231012.1) and *Glomerella cingulata* (JX139533.1) with an out group using *Phytophthora capsici* (FJ535572.1) that separates the clades of the NJ tree (Figure 22).

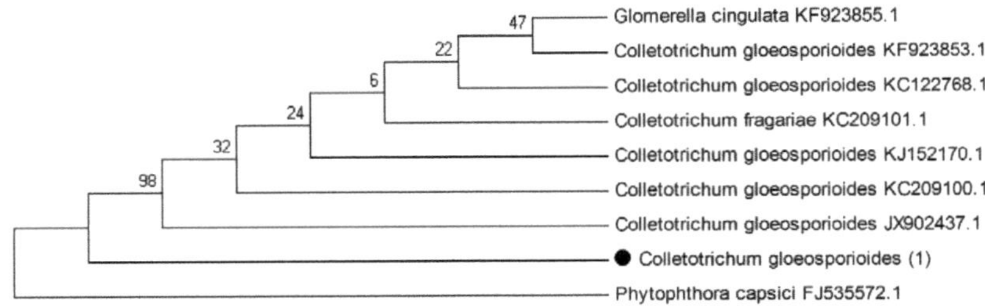

Figure 18. Phylogenetic tree of various *Colletotrichum* species based on DNA sequences. Values from 1000 bootstrap replications are shown above tree branches. Code after the species names are the NCBI BLAST accession number

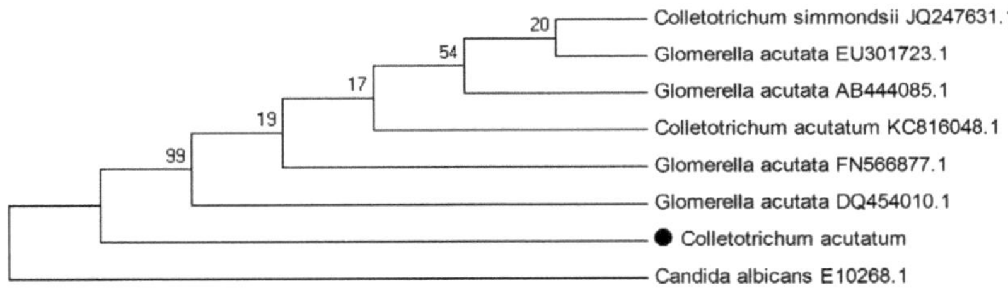

Figure 19. Phylogenetic tree of various *Colletotrichum* species based on DNA sequences. Values from 1000 bootstrap replications are shown above tree branches. Code after the species names are the NCBI BLAST accession number

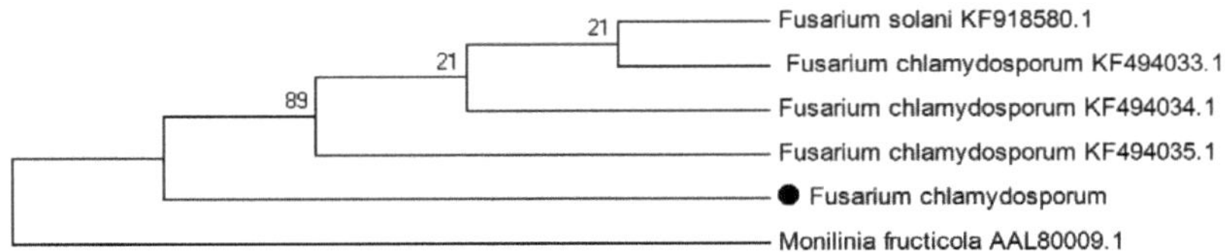

Figure 20. Phylogenetic tree of various *Fusarium* species based on DNA sequences. Values from 1000 bootstrap replications are shown above tree branches. Code after the species names are the NCBI BLAST accession number

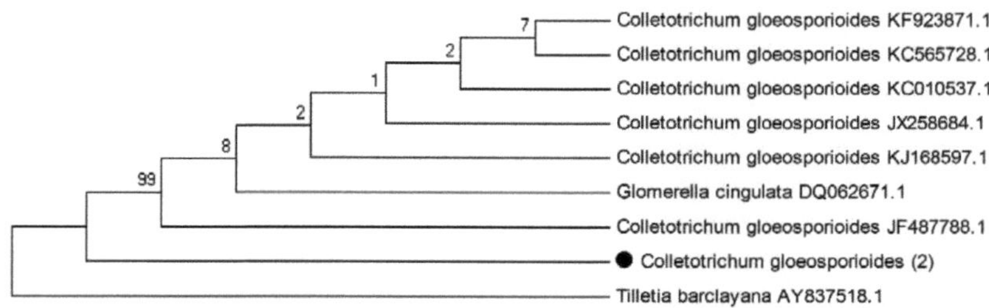

Figure 21. Phylogenetic tree of various *Colletotrichum* 2 species based on DNA sequences. Values from 1000 bootstrap replications are shown above tree branches. Code after the species names are the NCBI BLAST accession number

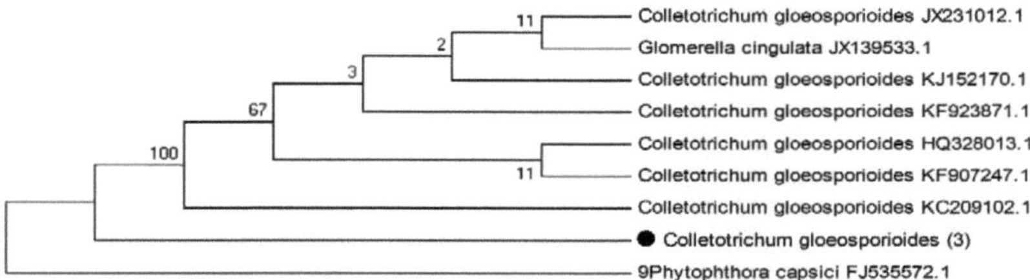

Figure 22. Phylogenetic tree of various *Colletotrichum* 3 species based on DNA sequences. Values from 1000 bootstrap replications are shown above tree branches. Code after the species names are the NCBI BLAST accession number

A definite identification of *Colletotrichum* species based on morphology is difficult because isolates have overlapping ranges of conidial and colony characteristics, and because the variation in morphology is accepted for isolates within a species (Sutton, 1992). *Fusarium* species emphasizes the need for accurate identification on the species level. According to Leslie and Summerell (2006), *Fusarium* species can be identified based on morphology alone. However, identification based on molecular data is considered more reliable and accurate than morphological identification and has become much more important in diagnostics of the fungi from the genus *Fusarium*.

Based from the results, molecular analyses indicated that *Colletotrichum* species from GenBank was phylogenetically related to the *Colletotrichum* species isolated from guyabano fruit and leaves. According to Johnston and Jones (1997), *Colletotrichum acutatum, Fusarium.* sp. and *Glomerella miyabeana*, the pathogens of willow, were found to be related phylogenetically to *Colletotrichum acutatum* when using the D1–D2 LSU rDNA region.

Alvarez et al. (2005) also confirms that *Colletotrichum gloeosporioides* and *Colletotrichum acutatum* are associated with anthracnose in soursop.

SUMMARY, CONCLUSIONS AND RECOMMENDATIONS

Pathogens attacking guyabano fruits and leaves in Muñoz were characterized and identified morphologically and molecularly. The five isolated pathogens from the infected guyabano fruit and leaves were identified as *Colletotrichum gloeosporioides* 1, *Colletotrichum acutatum*, *Fusarium chlamydosporum*, *Colletotrichum gloeosporioides* 2 and *Colletotrichum gloeosporioides* 3. The pathogens sensitivity to different fungicides was also determined. Two different factors in the fungicide sensitivity were the Fungicides as Factor 1 and Rate of Fungicide as Factor 2. The Factor 1 are as follows: Tebuconazole, Chlorothalonil (1), Propineb, Benomyl, Cupric hydroxide, Mancozeb, Chlorothalonil (2), Difeconazole + Propiconazole, Captan, Azoxystrobin Carbendazim and Thiophanate Methyl and Factor 2 are as follows: Rate 1- Recommended rate, Rate II- Recommended rate – 25% and Rate III- Recommended rate +25%.

Colletotrichum gloeosporioides 1 was found to be sensitive to DP in all concentrations in the sensitivity test, as well as Tebuconazole in recommended rate and recommended rate +25%. Captan and Difeconazole + Propiconazole were also found to be effective against *Colletotrichum acutatum* in all rates and *Fusarium chlamydoporum* to Difeconazole + Propiconazole and Captan in all rates tested. *Colletotrichum gloeosporioides* 2 and *Colletotrichum gloeosporioide* 3 were also found to be sensitive to Captan and Difeconazole + Propiconazole. Different rates of fungicides tested were found to have different degrees of efficacy against all isolated pathogens. For final validation of the results, Pot and field trials on the tested fungicides must be conducted in the future.

PCR amplification of species-specific designed primer CgInt2 and CaInt and universal primer Ver ITS4 gives uniform amplification in all the isolates of

Colletotrichum gloeosporioides, Colletotrichum acutatum and *Fusarium chlamydosporum.* Thus, the designed primer proved to be an efficient marker for species-specific discrimination which would be useful in developing a rapid and sensitive diagnostic PCR based assay for early detection and timely management of diseases.

Identification of the causal agent and prevalence of a disease is very essential for adequate and timely management of disease, which in turn depends also on accurate diagnosis and early detection of the pathogen. Therefore, the use of accurate and rapid identification techniques for early detection and identification of these pathogens is needed. Additional species-specific PCR assays and DNA sequence analyses are desirable to facilitate accurate identifications.

LITERATURE CITED

ADISA, V.A. and A.O. FAJOLA. 1982. Post- harvest fruit rots of three species of citrus in Southern Nigeria. Indian Phytopathol. 35:595-603.

ALVAREZ, E., J.F. MEJIA, G. LLANO and J. LOKE, 2005. Characterization of *Colletotrichum gloeosporioides*, causal agent of anthracnose in soursop (*Annona muricata*) in Valle del Cauca, Colombia. International Center for Tropical Agriculture (CIAT).28:1-2.

AMUSA, N.A., O.A. ASHAYE, M.O. OLADAPO and O.O. KAFARU. 2003. Pre-harvest deterioration of Soursop (*Annona muricata*) at Ibadan Southwestern Nigeria and its effect on nutrient composition. African Journal of Biotechnology. 2(1):23-25.

AGRIOS, G. 2005. Plant Pathology.4th edition. Academic press, New York. 358 pp.

BAXTER, A.P., G.C.A VAN DER WESTHUIZEN and A. EICKER. 1983: Morphology and taxonomy of South African isolates of *Colletotrichum*. South African Journal of Botany, 2: 259-289.

BILAI, V.I. 1988. Agents of Plant Diseases. 3rd ed. Kiev: Naukovadumka. 550 pp.

DE Q. PINTO, A.C., MC.R. CORDEIRO, S.R.M. DE ANDRADE, F.R. FERREIRA, H.A. FILGUEIRAS, R.E. ALVES and D.I. KINPARA. 2005. Fruits for the Future 5- annona species. The International Centre for Underutilised Crops, University of Southampon, Southampon, UK.82 pp.

DAHMEN, H. and T. STAUB.1992. Protective, curative and eradicant activity of difeconazole against *Venturia inaequalis Cercospora arachidicola*, and *Alternaria solani*.Plant Dis. 76:774-777

DEACON, J.W. 2006. Fungal Biology, 4th edition. Blackwell Publishing Ltd. pp. 205.pdf

DOUGLAS, S.M. 2011. Anthracnose diseases of trees. Department of Plant Pathology and Ecology, The Connection Agricultural Experiment Station. Retrieved on April 14, 2013 at http://www.ct/govcaes.

DICKMAN, M.B. 1993. Latent infection of papaya caused by *Colletotrichum gloeoesporioides*: Plant Dis. 67:748-750.

DYKO, B.A. and J.E.M, MORDUE. 1979. CMI descriptions of pathogenic fungi and bacteria. No. 630, *Colletotrichum acutatum*. Commonwealth Mycological Institute, Kew, Surrey, England. 26:352-357.

EVERETT, K. and O. THOMAS. 2001. Plant Pathology. 3rdedition.Wiley and Sons, New York. 465 pp.

GOTTSBERGER, G. 1988. Comments on flower evolution and beetle pollination in the genera *Annona* and *Rollinia*. Plant Systematics and Evolution. 167 (3-4): 189-194.

HOOPER, J., B. BONEV AND J. PARISOT. 2008. Principles of assessing bacterial susceptibility to antibiotics using the agar diffusion method.61:6: 1295-1301 pp.

HOSSAIN, M. T., S. M. M. HOSSAIN, M. A. BAKR A. K. M. MATIAR RAHMAN and S. N. UDDIN. 2010. Survey on major diseases of vegetable and fruit crops in Chittagong Region. Res. 35(3): 423-429

HOSSEIN, G. 2011. *Colletotrichum gloeosporioides*. Pathogen of the Month. Retrieved on May 6, 2013 at http://www.agric.wa.gov.

JARAMILLO, M., N.C. MONGHATA and P. PETER, 2000. Cytotoxicity and antileishmanial activity of *Annona muricata* pericarp. Fitoterapia. 71: 183-186.

JOHNSTON, P.R and D. JONES. 1997. Relationships among *Colletotrichum* isolates from fruit-rots assessed using rDNA sequences. Mycologia 89:420 - 430.

KOMANSILAN, A., A. L. ABADI, B. YANUWIADI, D.A. KALIGIS. 2012. Isolation and identification of biolarvicide from soursop (*Annona muricata* Linn) seeds to mosquito (Aedesaegypti) larvae. International Journal of Engineering &Technology IJET-IJENS. 12(03):28-32.

KUNZ, R. 2007. Control of post-harvest disease (*Botryodiplodi*a sp.) of Rambutan and Annona species by using bio-control agent (*Trichoderma* sp.). The Internationational Centre Underutilised Crops. 16-17 pp.

LESLIE J.F. and B.A. SUMMERELL. 2006 The *Fusarium* Laboratory Manual. Ames (IA): Blackwell Publishing. 63:193-213.

LOVE, K. and R. E. PAULL. 2011. Soursop.hawaii tropical fruit growers, CTAHR Department of Tropical Plant and Soil Sciences. 5 pp.

LUNN, J. A. 2004. Pathogenic Fungi and Bacteria.2nd ed. A.P.S. Press. 524 pp.

McDONALD, D. and A.Z. JOFFE. 1990. Fungal infection of groundnut fruit after maturity and during drying. Trans Br Mycol Soc.54:461–472.

MORTON, J. 1987. Soursop. In: fruits of Warm Climates. Julia.F. and Miami, F.L. (eds).   76 pp.

MORTON, J. 2000. Fruits of Warm Climates. 2nd ed. University of Florida. 401 pp.

MUELLER, D. S. and C.A. BRADLEY. 2010. Field Crop Fungicides for the North Central United States. North Central IPM Center. University of Illinois. 17 pp.

NWEKE, C.N. and O.F.A. IBIAM. 2012. Pre and post-harvest fungi associated with the soft rot of the fruit of *Annona muricata*, and their effects on the nutrient content of the pulp. Department of Applied Biology, Faculty of Biological Sciences, Ebonyi State University, Abakaliki.Nigeria.2(4):78-85.

OKIGBO, R. N. and O. OBIRE. 2009. Mycroflora and production of wine from fruits of soursop (*Annon amuricata*). Department of Botany, Nnamdi AzikiweUniversity P.M.B. 5025, Awka, Anambra State, Nigeria.1;1-9.

ONYEANI, C. A., S, OSUNLAJA, O. O. OWORU., S, OLUFEMI. 2012. First Report of Fruit Anthracnose in Mango caused by *Colletotrichum gloeosporioides* in Southwestern Nigeria. 1:1-7

OU S. H. 1985.   Rice Diseases, 2nd edition. The Cambrian News (Aberystwyth) Ltd.380p.

PITT, J.I, and A.D. HOCKING. 2009. Fungi and Food Spoilage, 3rd edition. Springer Dordrecht Heidelberg London New York. 4: 53-353.pdf

PLOETZ, R.C. 2003. 2 Diseases of Atemoya, Cherimoya, Soursop, Sugar apple and Related Fruit crops. (EDS) R.C. Ploetz, CAP International. 26 pp.

RITCHIE, D.F. 2000. Brown rot of stone fruits. The Plant Health Instructor. DOI: 10.1094/PHI-I-2000-1025-01

SETIAWAN, B., A. SULAEMAN and D.W. GIRAUD. 2001. Carotenoid content of selected Indonesians fruits. J. Food. Comp. Anal. 14: 169-176.

SINGH, S. P. 1992. Fruit Crops for Wasteland. Scientific Publisher. Jodhpur, India. 35 pp.

SMITH, B.J. and L.L. BLACK. 1990. Morphological, cultural, and pathogenic variation among *Colletotrichum* species isolated from strawberry. Plant Disease, 4: 69-76.

SUTTON, B.C. 1992: The genus *Glomerella* and its anamorph *Colletotrichum* In: *Colletotrichum*: Biology, Pathology and Control (Bailey J.A., Jeger M.J., eds.), CAB Int., Wallingford, UK, 26 pp.

TAMURA, K., J. DUDLEY, M. NEI and S. KUMAR. 2007. MEGA4: Molecular Evolutionary Genetics Analysis (MEGA) software version 4.0. Mol. Biol. Evol., 24: 1596- 1599.

TABASUM, Z. A. 2012. Soursop or Guyabano: cancer-killer fruit. LUMHS online clinical discussion forum. LUMHS online article. Retrieved on July 3, 2013 at www.alagad.com.ph/.../609-sour-sop-or-guyabano-cancer-killer-fruit.pdf.

WEBSTER, J. and R.W.S WEBER. 2007. Introduction to Fungi, 3rd edition. Cambridge University Press. 475 pp

WEDGE, D.E. 2005. Agrochemical Discovery: Finding new fungicides from natural products 3:1-5

http://infonewssoftware.blogspot.com.html retrieved on March 6, 2013.

http://www.blast.ncbi.nlm.gov retrieved on July 14, 2013.

http://www.ebi.ac.uk/Tools/clustal2 retrieved on July 14, 2013.

http://www. bpi.da.gov.ph retrieved on July 14, 2013.

http://www.echonet.org retrieved on May 3, 2013

APPENDICES

Appendix Table 1. Diameter (mm) of the zone of inhibition of the 12 fungicides tested against *Colletotrichum gloeosporioides* in guyabano fruit three days after incubation

FUNGICIDE	RECOMMENDED RATE			RECOMMENDED RATE - 25%			RECOMMENDED RATE + 25%		
	REPLICATES								
	I	II	III	I	II	III	I	II	III
1. Tebuconazole	14.44	14.25	9.13	10.25	1.63	9.13	11.38	8.63	9.50
2. Chlorothalonil 1	3.38	0.00	0.00	2.94	5.44	3.31	3.44	8.63	2.00
3. Propineb	8.06	4.00	6.00	2.75	4.25	2.50	8.75	12.44	10.25
4. Benomyl	2.44	2.19	1.88	7.00	9.88	4.94	3.63	7.63	4.94
5. Cupric Hydroxide	0.00	0.00	0.00	0.00	0.00	0.00	0.00	0.00	0.00
6. Mancozeb	3.00	4.31	2.69	3.38	4.31	2.06	2.94	1.63	3.63
7. Chlorothalonil 2	7.06	9.75	6.56	6.19	3.25	3.38	8.38	3.25	5.38
8. DP	19.75	20.88	22.00	17.50	20.31	22.00	21.88	20.94	20.19
9. Captan	13.81	13.63	16.31	13.00	8.64	16.94	13.81	12.88	11.50
10. Azoxystrobin	3.25	2.38	0.00	0.88	0.00	0.00	2.63	0.25	0.00
11. Carbendazim	0.00	0.00	0.81	0.63	1.25	1.13	0.00	0.00	2.34
12. Thiophanate Methyl	0.00	0.00	0.00	0.00	0.00	0.00	0.00	0.00	0.00
13. Control	0.00	0.00	0.00	0.00	0.00	0.00	0.00	0.00	0.00

DP= Difeconazole + Propiconazole

Appendix Table 2. Diameter (mm) of the zone of inhibition of the 12 fungicides tested against *Colletotrichum acutatum* in guyabano fruit three days after incubation

FUNGICIDE	RECOMMENDED RATE			RECOMMENDED RATE - 25%			RECOMMENDED RATE + 25%		
	REPLICATES								
	I	II	III	I	II	III	I	II	III
1. Tebuconazole	1.25	0.00	0.00	0.50	0.00	0.00	2.13	0.00	0.00
2. Chlorothalonil 1	2.63	4.63	3.88	3.25	1.75	5.38	8.75	6.00	5.75
3. Propineb	0.50	2.75	0.00	0.75	5.38	0.00	0.00	0.00	0.00
4. Benomyl	0.00	0.00	0.00	0.00	2.13	1.00	0.00	0.00	0.00
5. Cupric Hydroxide	0.00	0.00	0.00	0.00	0.00	0.00	0.00	0.00	0.00
6. Mancozeb	0.00	1.50	3.00	0.00	0.00	0.00	3.00	3.44	2.75
7. Chlorothalonil 2	3.88	3.25	2.31	0.75	2.38	1.13	3.25	3.63	2.25
8. DP	12.88	14.75	16.13	15.38	17.00	15.88	14.13	15.38	15.5
9. Captan	18.50	17.75	13.75	15.50	13.13	15.00	19.63	19.94	16.88
10. Azoxystrobin	8.13	6.50	7.875	3.50	1.50	0.88	2.38	4.88	3.88
11. Carbendazim	1.38	0.38	0.00	0.25	0.63	0.00	0.00	0.75	0.00
12. Thiophanate Methyl	1.50	0.00	0.00	1.00	0.00	0.00	2.00	0.00	0.00
13. Control	0.00	0.00	0.00	0.00	0.00	0.00	0.00	0.00	0.00

DP= Difeconazole + Propiconazole

Appendix Table 3. Diameter (mm) of the zone of inhibition of the 12 fungicides tested against *Fusarium sp.* in guyabano fruit three days after incubation

FUNGICIDE	RECOMMENDED RATE			RECOMMENDED RATE - 25%			RECOMMENDED RATE + 25%		
	REPLICATES								
	I	II	III	I	II	III	I	II	III
1. Tebuconazole	0.00	0.00	0.00	0.00	0.00	0.00	0.00	0.00	0.00
2. Chlorothalonil 1	3.19	2.38	1.00	0.00	3.50	0.00	0.00	0.00	0.00
3. Propineb	6.00	6.63	9.25	6.88	5.63	6.50	0.00	1.13	7.13
4. Benomyl	0.00	0.00	0.00	0.00	0.00	0.00	0.00	0.00	0.00
5. Cupric Hydroxide	0.00	0.00	0.00	0.00	0.00	0.00	0.00	0.00	0.00
6. Mancozeb	0.38	0.00	0.00	0.81	1.00	0.00	6.31	0.63	1.50
7. Chlorothalonil 2	1.00	1.50	0.50	0.00	0.00	1.75	1.06	0.00	0.00
8. DP	15.75	17.00	16.00	19.19	17.63	17.88	18.13	15.50	14.38
9. Captan	17.13	19.94	16.88	7.75	9.00	15.31	16.25	16.63	15.50
10. Azoxystrobin	0.63	5.25	3.75	8.19	13.94	1.50	6.50	11.06	9.88
11. Carbendazim	6.81	4.06	4.56	3.81	6.06	3.44	7.25	6.81	7.56
12. Thiophanate Methyl	0.00	0.00	0.00	0.00	0.00	0.00	0.00	0.00	0.00
13. Control	0.00	0.00	0.00	0.00	0.00	0.00	0.00	0.00	0.00

DP= Difeconazole + Propiconazole

Appendix Table 4. Diameter (mm) of the zone of inhibition of the 12 fungicides tested against *Colletotrichum gloeosporioides* in guyabano leaves from Magtanggol three days after incubation

FUNGICIDE	RECOMMENDED RATE			RECOMMENDED RATE - 25%			RECOMMENDED RATE + 25%		
	REPLICATES								
	I	II	III	I	II	III	I	II	III
1. Tebuconazole	2.44	0.81	1.38	3.44	5.75	4.63	0.25	0.75	2.13
2. Chlorothalonil 1	4.25	2.81	3.88	3.13	1.75	5.25	1.81	1.75	0.00
3. Propineb	4.50	0.75	1.00	2.38	2.19	3.188	3.63	4.38	4.06
4. Benomyl	0.63	2.00	1.00	0.69	0.00	1.19	1.44	1.94	0.25
5. Cupric Hydroxide	1.63	0.00	0.00	1.25	1.38	0.00	0.00	1.00	0.00
6. Mancozeb	0.00	0.00	0.00	3.06	3.94	2.81	2.50	4.81	3.25
7. Chlorothalonil 2	3.13	3.00	3.75	3.69	1.00	4.81	4.56	4.13	5.38
8. DP	19.00	14.25	17.50	15.38	13.63	15.75	17.75	21.19	21.13
9. Captan	16.06	13.75	15.63	14.88	14.50	12.13	19.88	19.88	19.13
10. Azoxystrobin	4.31	3.31	2.19	1.13	3.13	2.88	6.34	1.25	2.00
11. Carbendazim	0.00	1.25	1.13	0.50	0.00	0.31	0.00	0.44	0.75
12. Thiophanate Methyl	2.63	6.44	5.19	10.38	7.63	6.63	0.00	0.00	0.00
13. Control	0.00	0.00	0.00	0.00	0.00	0.00	0.00	0.00	0.00

DP= Difeconazole + Propiconazole

Appendix Table 5. Diameter (mm) of the zone of inhibition of the 12 fungicides tested against *Colletotrichum gloeosporioides* in guyabano leaves from Catalanacan three days after incubation

FUNGICIDE	RECOMMENDED RATE			RECOMMENDED RATE - 25%			RECOMMENDED RATE + 25%		
	REPLICATES								
	I	II	III	I	II	III	I	II	III
1. Tebuconazole	4.81	1.75	2.94	0.75	2.44	1.50	7.00	6.50	3.31
2. Chlorothalonil 1	0.63	3.06	1.25	0.00	0.00	0.00	0.81	1.81	5.63
3. Propineb	1.38	2.75	2.50	3.63	1.44	2.38	5.00	2.50	2.69
4. Benomyl	7.31	5.56	6.50	0.50	0.00	0.00	2.88	0.75	5.00
5. Cupric Hydroxide	0.00	0.00	0.00	0.00	0.00	0.00	1.25	1.25	1.25
6. Mancozeb	1.13	1.56	0.00	0.00	0.00	1.38	1.63	1.63	0.94
7. Chlorothalonil 2	0.00	3.75	6.19	3.88	4.75	4.94	3.38	2.63	1.25
8. DP	21.63	22.75	16.13	20.13	21.63	17.44	21.44	20.56	18.88
9. Captan	17.13	14.94	17.69	15.88	14.50	13.88	18.00	16.56	19.75
10. Azoxystrobin	0.00	0.00	0.31	5.63	4.69	2.69	4.38	2.00	2.75
11. Carbendazim	0.00	0.88	1.00	0.38	0.00	0.88	0.00	0.50	1.36
12. Thiophanate Methyl	0.00	0.00	0.00	0.00	0.00	0.00	0.00	0.00	0.00
13. Control	0.00	0.00	0.00	0.00	0.00	0.00	0.00	0.00	0.00

DP= Difeconazole + Propiconazole

Appendix Table 6. Analysis of variance on disease incidence (%) on the guyabano fruits three days after inoculation

SOURCE OF VARIATION	SUM OF SQUARES	DEGREES OF FREEDOM	MEAN SQUARE	F VALUE	P-VALUE
Model	22500.000	3	7500.0000	Infty	<0.0001**
Error	0.000	8	0.0000		
Total	22500.000	11			

** = highly significant at 1% level

Appendix Table 7. Analysis of variance on disease severity (%) on the guyabano fruits three days after inoculation

SOURCE OF VARIATION	SUM OF SQUARES	DEGREES OF FREEDOM	MEAN SQUARE	F VALUE	P-VALUE
Model	16679.630	3	5559.8769	187.70	0.0000**
Error	236.9630	8	29.6204		
Total	16916.593	11	1537.8722		

** = highly significant at 1% level

Appendix Table 8. Analysis of variance on disease incidence (%) on the guyabano leaves three days after inoculation

SOURCE OF VARIATION	SUM OF SQUARES	DEGREES OF FREEDOM	MEAN SQUARE	F VALUE	P-VALUE
Model	20000.00000	2	10000.000	Infty	<0.0001**
Error	0.00000	6	0.000		
Total	20000.00000	8			

** = highly significant at 1% level

Appendix Table 9. Analysis of variance on disease severity (%) on the guyabano leaves three days after inoculation

SOURCE OF VARIATION	SUM OF SQUARES	DEGREES OF FREEDOM	MEAN SQUARE	F VALUE	P-VALUE
Model	9956.5923	2	4978.2962	99.29	<0.0002**
Error	300.8452	6	50.1409		
Total	10257.4375	8	1282.1797		

** = highly significant at 1% level

Appendix Table 10. Analysis of variance on the diameter (mm) of the zone of inhibition of the 12 fungicides tested against *Colletotrichum gloeosporioides* in guyabano fruit from Magtanggol three days after incubation

SOURCE OF VARIATION	SUM OF SQUARES	DEGREES OF FREEDOM	MEAN SQUARE	F VALUE	P-VALUE
Model	7639.364[a]	39	195.881	63.441	.000**
Fungicide	4093.702	12	341.142	110.4881	.000**
Rate	15.450	2	7.725	2.502	0.88ns
Fungicide Rate	206.745	24	8.614	2.79	.000**
Error	240.832	78	3.088		
Total	7880.19625	117			

** = highly significant at 1% level

ns = not significant

Appendix Table 11. Analysis of variance on the diameter (mm) of the zone of inhibition of the 12 fungicides tested against *Colletotrichum acutatum* in guyabano fruit from Villa Cuison three days after incubation

SOURCE OF VARIATION	SUM OF SQUARES	DEGREES OF FREEDOM	MEAN SQUARE	F VALUE	P-VALUE
Model	5180.837[a]	39	132.842	109.103	.000**
Fungicide	3476.271	12	289.689	237.922	.000**
Rate	14.700	2	7.350	6.036	0.04**
Fungicide Rate	114.574	24	4.774	3.921	.000**
Error	94.971	78	3.921		
Total	5275.809	117			

** = highly significant at 1% level

Appendix Table 12. Analysis of variance on the diameter (mm) of the zone of inhibition of the 12 fungicides tested against *Fusarium chlamydosporum* in guyabano fruit three days after incubation

SOURCE OF VARIATION	SUM OF SQUARES	DEGREES OF FREEDOM	MEAN SQUARE	F VALUE	P-VALUE
Model	5762.055[a]	39	147.745	50.795	.000**
Fungicide	3633.134	12	302.761	104.090	.000**
Rate	2.479	2	1.239	0.426	0.655ns
Fungicide Rate	218.269	24	9.095	3.127	.000**
Error	226.875	78	2.909		
Total	5988.930	117			

** = highly significant at 1% level

ns = not significant

Appendix Table 13. Analysis of variance on the diameter (mm) of the zone of inhibition of the 12 fungicides tested against *Colletotrichum gloeosporioides* in guyabano fruit from Magtanggol three days after incubation

SOURCE OF VARIATION	SUM OF SQUARES	DEGREES OF FREEDOM	MEAN SQUARE	F VALUE	P-VALUE
Model	5943.560[a]	39	152.399	107.616	.000**
Fungicide	3439.574	12	286.631	202.403	.000**
Rate	4.788	2	2.394	1.690	0.191ns
Fungicide Rate	261.928	24	10.914	7.707	.000**
Error	110.459	78	1.416		
Total	6054.019	117			

** = highly significant at 1% level

ns = not significant

Appendix Table 14. Analysis of variance on the diameter (mm) of the zone of inhibition of the 12 fungicides tested against *Colletotrichum gloeosporioides* in guyabano fruit from Catalanacan three days after incubation

SOURCE OF VARIATION	SUM OF SQUARES	DEGREES OF FREEDOM	MEAN SQUARE	F VALUE	P-VALUE
Model	6676.474[a]	39	171.192	101.103	.000**
Fungicide	4418.495	12	368.208	217.457	.000**
Rate	20.514	2	10.257	6.058	0.004**
Fungicide Rate	134.764	24	5.615	3.316	.000**
Error	132.073	78	1.693		
Total	6808.547	117			

** = highly significant at 1% level

Printed by Books on Demand GmbH, Norderstedt / Germany